FAREWELL SPEECHES

First Melville House Printing: March 2017

Melville House Publishing
46 John Street
Brooklyn, NY 11201

and

8 Blackstock Mews
Islington
London N4 2BT

mhpbooks.com   facebook.com/mhpbooks   @melvillehouse

ISBN: 978-1-61219-663-3

Printed in the United States of America
1 3 5 7 9 10 8 6 4 2

A catalog record for this book is available from the Library of Congress

FAREWELL SPEECHES

BARACK & MICHELLE OBAMA

MELVILLE HOUSE
BROOKLYN · LONDON

Contents

MICHELLE OBAMA'S FINAL ADDRESS

JANUARY 6, 2017

THE WHITE HOUSE, WASHINGTON, D.C.

Michelle Obama's Final Address

January 6, 2017

Hello, everyone. And, may I say for the last time officially, welcome to the White House. Yes! *(Applause.)* Well, we are beyond thrilled to have you all here to celebrate the 2017 National School Counselor of the Year, as well as all of our State Counselors of the Year. These are the fine women, and a few good men—*(laughter)*—one good man—who are on this stage, and they represent schools from across this country.

And I want to start by thanking Terri for that wonderful introduction and her right-on-the-spot remarks. I'm going to say a lot more about Terri in a few minutes, but first I want to take a moment to acknowledge a few people who are here.

First, our outstanding Secretary of Education, John King. *(Applause.)* As well as our former Education Secretary, Arne Duncan. *(Applause.)* I want to take this time to thank you both publicly for your dedication and leadership and friendship. We couldn't do this without the support of the Department of Education under both of your leadership. So I'm grateful to you personally, and very proud of all that you've done for this country.

I also want to acknowledge a few other special guests we have in the audience. We've got a pretty awesome crew. As one of my staff said, "You roll pretty deep." *(Laughter.)* I'm like, well, yeah, we have a few good friends. We have with us today Ted Allen, La La Anthony, Connie Britton, Andy Cohen—yeah, Andy Cohen is here—*(laughter)*—Carla Hall, Coach Jim Harbaugh and his beautiful wife, who's a lot better looking than him—*(laughter)*—Lana Parrilla, my buddy Jay Pharoah, Kelly Rowland, Usher—

(Audience Member: Woo!)

Keep it down. *(Laughter.)* Keep it together, ladies. Wale is here. And of course, Allison Williams and her mom are here.

And all these folks are here because they're using their star power to inspire our young people. And I'm so grateful to all of you for stepping up in so many ways on so many occasions. I feel like I've pestered you over these years, asking time and time again, "Well, where are you going to be?" "I'm going to be in New York." "Can you come? Can you come here? Can you do this? Can you take that? Can you ask for that? Can you come? Can we rap? Can we sing?" *(Laughter.)* So thank you all so much. It really means the world to this initiative to have such powerful, respected and admired individuals speaking on behalf of this issue. So congratulations on the work that you've done, and we're going to keep working.

And today, I especially want to recognize all these—extraordinary leadership team that was behind Reach Higher from day one. And this isn't on the script so they don't know this. I want to take time to personally acknowledge a couple of people. Executive Director Eric Waldo. *(Applause.)* Where is Eric? He's in the— you've got to step out. *(Applause.)* Eric is acting like he's a ham, but he likes the spotlight. *(Laughter.)* He's acting a little shy.

I want to recognize our Deputy Director, Stephanie Sprow. Stephanie. *(Applause.)* And he's really not going to like this because he tries to pretend like he doesn't exist at all, but our Senior Advisor, Greg Darnieder. *(Applause.)* There you go. Greg has been a leader in education his entire life. I've known him since I was a little organizer person. And it's just been just a joy to work with you all. These individuals, they are brilliant. They are creative. They have worked miracles with hardly any staff or budget to speak of— which is how we roll in the First Lady's Office. *(Laughter.)* And I am so proud and so, so grateful to you all for everything that you've done. So let's give them a round of applause. *(Applause.)*

And finally, I want to recognize all of you who are here in this audience. We have our educators, our leaders, our young people who have been with us since we launched Reach Higher back in 2014.

Now, when we first came up with this idea, we had one clear goal in mind: We wanted to make higher education cool. We wanted to change the conversation around what it means and what it takes to be a success in this country. Because let's be honest, if we're always shining the spotlight on professional athletes or recording artists or Hollywood celebrities, if those are the only achievements we celebrate, then why would we ever think kids would see college as a priority?

So we decided to flip the script and shine a big, bright spotlight on all things educational. For example, we made College Signing Day a national event. We wanted to mimic all the drama and excitement traditionally reserved for those few amazing football and basketball players choosing their college and university teams. We wanted to focus that same level of energy and attention on kids going to college because of their academic achievements. Because as a nation, that's where the spotlight should also be—on kids who work hard in school and do the right thing when no one is watching, many beating daunting odds.

Next, we launched Better Make Room. It's a social media campaign to give young people the support and inspiration they need to actually complete higher education. And to really drive that message home, you may recall that I debuted my music career—*(laughter)*—rapping with Jay about getting some knowledge by going to college. *(Laughter and applause.)*

We are also very proud of all that this administration has done to make higher education more affordable. We doubled investments in Pell grants and college tax credits. We expanded income-based loan repayment options for tens of millions of students. We made it easier to apply for financial aid. We created a College Scorecard to help students make good decisions about higher education. And we provided new funding and support for school counselors. (Applause.) Altogether, we made in this administration the largest investment in higher education since the G.I. Bill. *(Applause.)* And today, the high school graduation rate is at a record high, and more young people than ever before are going to college.

And we know that school counselors like all of the folks standing with me on this stage have played a critical role in helping us get there. In fact, a recent study showed that students who met with a school counselor to talk about financial aid or college were three times more likely to attend college, and they were nearly seven times more likely to apply for financial aid.

So our school counselors are truly among the heroes of the Reach Higher story. And that's why we created this event two years ago, because we thought that they should finally get some recognition. *(Applause.)* We wanted everyone to know about the difference that these phenomenal men and women have been making in the lives of our young people every day. And our 2017 School Counselor of the Year, Terri Tchorzynski, is a perfect example.

As you heard, Terri works at the Calhoun Area Career Center, a career and technical education school in Michigan. And here's what Terri's principal said about her in his letter of recommendation. He said, "Once she identifies a systemic need, she works tirelessly to address it."

So when students at Terri's school reported feeling unprepared to apply for higher education, Terri sprang into action to create a school-wide, top-to-bottom college-readiness effort. Under Terri's leadership, more students than ever before attended workshops on resume writing, FAFSA completion—yes, I can now say FAFSA—*(laughter)*—and interview preparation. I can barely say it. *(Laughter.)* They did career and personal—personality assessments. They helped plan a special college week. And they organized a Military Day, hosting recruiters from all branches of our armed forces. And because of these efforts, today, 75 percent of Calhoun's seniors now complete key college application steps, and Terri's school has won state and national recognition.

And all of this is just one small part of what Terri does for her students each day. I can go on and on about all the time she spends one-on-one with students, helping them figure out their life path. Terri told us—as you heard, she told us about one of those students, so we reached out to

Kyra. And here's what Kyra had to say in her own words. Kyra wrote that "Mrs. Tchorzynski has helped me grow to love myself. She helped me with my doubts and insecurities." She said, my life has changed "for the better in all aspects." Kyra said, "She held my hand through my hardest times." She said, "Mrs. Tchorzynski is my lifesaver." That's what Kyra said. *(Laughter.)*

And this is what each of you do every single day. You see the promise in each of your students. You believe in them even when they can't believe in themselves, and you work tirelessly to help them be who they were truly meant to be. And you do it all in the face of some overwhelming challenges—tight budgets, impossible student-counselor ratios—

yeah,   amen—*(laughter)*—endless demands on your time.

You all come in early, you stay late. You reach into your own pockets—and see, we've got the amen corner. *(Laughter.)* You stick with students in their darkest moments, when they're most anxious and afraid. And if anyone is dealing with a college [high school] senior or junior, you know what this feels like. These men and women show them that those kids matter; that they have something to offer; that no matter where they're from or how much money their parents have, no matter what they look like or who they love or how they worship or what language they speak at home, they have a place in this country.

And as I end my time in the White House, I can think of no better message to send our young people in my last official remarks as First Lady. So for all the young people in this room and those who are watching, know that this country belongs to you—to all of you, from every background and walk of life. If you or your parents are immigrants, know that you are part of a proud American tradition—the infusion of new cultures, talents and ideas, generation after generation, that has made us the greatest country on earth.

If your family doesn't have much money, I want you to remember that in this country, plenty of folks, including me and my husband—we started out with very little. But with a lot of hard work and a good education, anything is possible—even becoming President. That's what the American Dream is all about. *(Applause.)*

If you are a person of faith, know that religious diversity is a great American tradition, too. In fact, that's why people first came to this country—to worship freely. And whether you are Muslim, Christian, Jewish, Hindu, Sikh—these religions are teaching our young people about justice, and compassion, and honesty. So I want our young people to continue to learn and practice those values with pride. You see, our glorious diversity—our diversities of faiths and colors and creeds—that is not a threat to who we are, it makes us who we are. *(Applause.)* So the young people here and the young people out there: Do not ever let anyone make you feel like you don't matter, or like you don't have a place in our American story—because you do. And you have a right to be exactly who you are.

But I also want to be very clear: This right isn't just handed to you. No, this right has to be earned every single day. You cannot take your freedoms for granted. Just like generations who have come before you, you have to do your part to preserve and protect those freedoms. And that starts right now, when you're young.

Right now, you need to be preparing yourself to add your voice to our national conversation. You need to prepare yourself to be informed and engaged as a citizen, to serve and to lead, to stand up for our proud American values and to honor them in your daily lives. And that means getting the best education possible so you can think critically, so you can express yourself clearly, so you can get a good job and support yourself and your family, so you can be a positive force in your communities.

And when you encounter obstacles— because I guarantee you, you will, and many of you already have—when you are struggling and you start thinking about giving up, I want you to remember something that my husband and I have talked about since we first started this journey nearly a decade ago, something that has carried us through every moment in this White House and every moment of our lives, and that is the power of hope—the belief that something better is always possible if you're willing to work for it and fight for it.

It is our fundamental belief in the power of hope that has allowed us to rise above the voices of doubt and division, of anger and fear that we have faced in our own lives and in the life of this country. Our hope that if we work hard enough and believe in ourselves, then we can be whatever we dream, regardless of the limitations that others may place on us. The hope that when people see us for who we truly are, maybe, just maybe they, too, will be inspired to rise to their best possible selves.

That is the hope of students like Kyra who fight to discover their gifts and share them with the world. It's the hope of school counselors like Terri and all these folks up here who guide those students every step of the way, refusing to give up on even a single young person. Shoot, it's the hope of my— folks like my dad who got up every day to do his job at the city water plant; the hope that one day, his kids would go to college and have opportunities he never dreamed of.

That's the kind of hope that every single one of us—politicians, parents, preachers—all of us need to be providing for our young people. Because that is what moves this country forward every single day—our hope for the future and the hard work that hope inspires.

So that's my final message to young people as First Lady. It is simple. *(Applause.)* I want our young people to know that they matter, that they belong. So don't be afraid—you hear me, young people? Don't be afraid. Be focused. Be determined. Be hopeful. Be empowered. Empower yourselves with a good education, then get out there and use that education to build a country worthy of your boundless promise. Lead by example with hope, never fear. And know that I will be with you, rooting for you and working to support you for the rest of my life.

And that is true I know for every person who are here—is here today, and for educators and advocates all across this nation who get up every day and work their hearts out to lift up our young people. And I am so grateful to all of you for your passion and your dedication and all the hard work on behalf of our next generation. And I can think of no better way to end my time as First Lady than celebrating with all of you.

So I want to close today by simply saying thank you. Thank you for everything you do for our kids and for our country. Being your First Lady has been the greatest honor of my life, and I hope I've made you proud. *(Applause.)*

PRESIDENT OBAMA'S FAREWELL ADDRESS

JANUARY 10, 2017

MCCORMICK PLACE CONVENTION CENTER, CHICAGO, ILLINOIS

President Obama's Farewell Address

January 10, 2017

Hello, Chicago! (Applause.) It's good to be home! (Applause.) Thank you, everybody. Thank you. (Applause.) Thank you so much. Thank you. (Applause.) All right, everybody sit down. (Applause.) We're on live TV here. I've got to move. (Applause.) You can tell that I'm a lame duck because nobody is following instructions. (Laughter.) Everybody have a seat. (Applause.)

My fellow Americans — (applause) — Michelle and I have been so touched by all the well wishes that we've received over the past few weeks. But tonight, it's my turn to say thanks. (Applause.) Whether we have seen eye-to-eye or rarely agreed at all, my conversations with you, the American people, in living rooms and in schools, at farms, on factory floors, at diners and on distant military outposts — those conversations are what have kept me honest, and kept me inspired, and kept me going. And every day, I have learned from you. You made me a better President, and you made me a better man. (Applause.)

So I first came to Chicago when I was in my early 20s. And I was still trying to figure out who I was, still searching for a purpose in my life. And it was a neighborhood not far from here where I began working with church groups in the shadows of closed steel mills. It was on these streets where I witnessed the power of faith, and the quiet dignity of working people in the face of struggle and loss.

AUDIENCE: Four more years! Four more years! Four more years!

THE PRESIDENT: I can't do that.

AUDIENCE: Four more years! Four more years! Four more years!

THE PRESIDENT: This is where I learned that change only happens when ordinary people get involved and they get engaged, and they come together to demand it.

After eight years as your President, I still believe that. And it's not just my belief. It's the beating heart of our American idea — our bold experiment in self-government. It's the conviction that we are all created equal, endowed by our Creator with certain unalienable rights, among them life, liberty, and the pursuit of happiness. It's the insistence that these rights, while self-evident, have never been self-executing; that We, the People, through the instrument of our democracy, can form a more perfect union.

What a radical idea. A great gift that our Founders gave to us: The freedom to chase our individual dreams through our sweat and toil and imagination, and the imperative to strive together, as well, to achieve a common good, a greater good.

For 240 years, our nation's call to citizenship has given work and purpose to each new generation. It's what led patriots to choose republic over tyranny, pioneers to trek west, slaves to brave that makeshift railroad to freedom. It's what pulled immigrants and refugees across oceans and the Rio Grande. (Applause.) It's what pushed women to reach for the ballot. It's what powered workers to organize. It's why GIs gave their lives at Omaha Beach and Iwo Jima, Iraq and Afghanistan. And why men and women from Selma to Stonewall were prepared to give theirs, as well. (Applause.)

So that's what we mean when we say America is exceptional — not that our nation has been flawless from the start, but that we have shown the capacity to change and make life better for those who follow. Yes, our progress has been uneven. The work of democracy has always been hard. It's always been contentious. Sometimes it's been bloody. For every two steps forward, it often feels we take one step back. But the long sweep of America has been defined by forward motion, a constant widening of our founding creed to embrace all and not just some. (Applause.)

If I had told you eight years ago that America would reverse a great recession, reboot our auto industry, and unleash the longest stretch of job creation in our history — (applause) — if I had told you that we would open up a new chapter with the Cuban people, shut down Iran's nuclear weapons program without firing a shot, take out the mastermind of 9/11 — (applause) — if I had told you that we would win marriage equality, and secure the right to health insurance for another 20 million of our fellow citizens — (applause) — if I had told you all that, you might have said our sights were set a little too high. But that's what we did. (Applause.) That's what you did.

You were the change. You answered people's hopes, and because of you, by almost every measure, America is a better, stronger place than it was when we started. (Applause.)

In 10 days, the world will witness a hallmark of our democracy.

AUDIENCE: Nooo —

THE PRESIDENT: No, no, no, no, no — the peaceful transfer of power from one freely elected President to the next. (Applause.) I committed to President-elect Trump that my administration would ensure the smoothest possible transition, just as President Bush did for me. (Applause.) Because it's up to all of us to make sure our government can help us meet the many challenges we still face.

We have what we need to do so. We have everything we need to meet those challenges. After all, we remain the wealthiest, most powerful, and most respected nation on Earth. Our youth, our drive, our diversity and openness, our boundless capacity for risk and reinvention means that the future should be ours. But that potential will only be realized if our democracy works. Only if our politics better reflects the decency of our people. (Applause.) Only if all of us, regardless of party affiliation or particular interests, help restore the sense of common purpose that we so badly need right now.

That's what I want to focus on tonight: The state of our democracy. Understand, democracy does not require uniformity. Our founders argued. They quarreled. Eventually they compromised. They expected us to do the same. But they knew that democracy does require a basic sense of solidarity — the idea that for all our outward differences, we're all in this together; that we rise or fall as one. (Applause.)

There have been moments throughout our history that threatens that solidarity. And the beginning of this century has been one of those times. A shrinking world, growing inequality; demographic change and the specter of terrorism — these forces haven't just tested our security and our prosperity, but are testing our democracy, as well. And how we meet these challenges to our democracy will determine our ability to educate our kids, and create good jobs, and protect our homeland. In other words, it will determine our future.

To begin with, our democracy won't work without a sense that everyone has economic opportunity. And the good news is that today the economy is growing again. Wages, incomes, home values, and retirement accounts are all rising again. Poverty is falling again. (Applause.) The wealthy are paying a fairer share of taxes even as the stock market shatters records. The unemployment rate is near a 10-year low. The uninsured rate has never, ever been lower. (Applause.) Health care costs are rising at the slowest rate in 50 years. And I've said and I mean it — if anyone can put together a plan that is demonstrably better than the improvements we've made to our health care system and that covers as many people at less cost, I will publicly support it. (Applause.)

Because that, after all, is why we serve. Not to score points or take credit, but to make people's lives better. (Applause.)

But for all the real progress that we've made, we know it's not enough. Our economy doesn't work as well or grow as fast when a few prosper at the expense of a growing middle class and ladders for folks who want to get into the middle class. (Applause.) That's the economic argument. But stark inequality is also corrosive to our democratic ideal. While the top one percent has amassed a bigger share of wealth and income, too many families, in inner cities and in rural counties, have been left behind — the laid-off factory worker; the waitress or health care worker who's just barely getting by and struggling to pay the bills — convinced that the game is fixed against them, that their government only serves the interests of the powerful — that's a recipe for more cynicism and polarization in our politics.

But there are no quick fixes to this long-term trend. I agree, our trade should be fair and not just free. But the next wave of economic dislocations won't come from overseas. It will come from the relentless pace of automation that makes a lot of good, middle-class jobs obsolete.

And so we're going to have to forge a new social compact to guarantee all our kids the education they need — (applause) —

to give workers the power to unionize for better wages; to update the social safety net to reflect the way we live now, and make more reforms to the tax code so corporations and individuals who reap the most from this new economy don't avoid their obligations to the country that's made their very success possible. (Applause.)

We can argue about how to best achieve these goals. But we can't be complacent about the goals themselves. For if we don't create opportunity for all people, the disaffection and division that has stalled our progress will only sharpen in years to come.

There's a second threat to our democracy — and this one is as old as our nation itself. After my election, there was talk of a post-racial America. And such a vision, however well-intended, was never realistic. Race remains a potent and often divisive force in our society. Now, I've lived long enough to know that race relations are better than they were 10, or 20, or 30 years ago, no matter what some folks say. (Applause.) You can see it not just in statistics, you see it in the attitudes of young Americans across the political spectrum.

But we're not where we need to be. And all of us have more work to do. (Applause.) If every economic issue is framed as a struggle between a hardworking white middle class and an undeserving minority, then workers of all shades are going to be left fighting for scraps while the wealthy withdraw further into their private enclaves. (Applause.) If we're unwilling to invest in the children of immigrants, just because they don't look like us, we will diminish the prospects of our own children — because those brown kids will represent a larger and larger share of America's workforce. (Applause.) And we have shown that our economy doesn't have to be a zero-sum game. Last year, incomes rose for all races, all age groups, for men and for women.

So if we're going to be serious about race going forward, we need to uphold laws against discrimination — in hiring, and in housing, and in education, and in the criminal justice system. (Applause.) That is what our Constitution and our highest ideals require. (Applause.)

But laws alone won't be enough. Hearts must change. It won't change overnight. Social attitudes oftentimes take generations to change. But if our democracy is to work in this increasingly diverse nation, then each one of us need to try to heed the advice of a great character in American fiction — Atticus Finch — (applause) — who said "You never really understand a person until you consider things from his point of view...until you climb into his skin and walk around in it."

For blacks and other minority groups, it means tying our own very real struggles for justice to the challenges that a lot of people in this country face — not only the refugee, or the immigrant, or the rural poor, or the transgender American, but also the middle-aged white guy who, from the outside, may seem like he's got advantages, but has seen his world upended by economic and cultural and technological change. We have to pay attention, and listen. (Applause.)

For white Americans, it means acknowledging that the effects of slavery and Jim Crow didn't suddenly vanish in the '60s — (applause) — that when minority groups voice discontent, they're not just engaging in reverse racism or practicing political correctness. When they wage peaceful protest, they're not demanding special treatment but the equal treatment that our Founders promised. (Applause.)

For native-born Americans, it means reminding ourselves that the stereotypes about immigrants today were said, almost word for word, about the Irish, and Italians, and Poles — who it was said we're going to destroy the fundamental character of America. And as it turned out, America wasn't weakened by the presence of these newcomers; these newcomers embraced this nation's creed, and this nation was strengthened. (Applause.)

So regardless of the station that we occupy, we all have to try harder. We all have to start with the premise that each of our fellow citizens loves this country just as much as we do; that they value hard work and family just like we do; that their children are just as curious and hopeful and worthy of love as our own. (Applause.)

And that's not easy to do. For too many of us, it's become safer to retreat into our own bubbles, whether in our neighborhoods or on college campuses, or places of worship, or especially our social media feeds, surrounded by people who look like us and share the same political outlook and never challenge our assumptions. The rise of naked partisanship, and increasing economic and regional stratification, the splintering of our media into a channel for every taste — all this makes this great sorting seem natural, even inevitable. And increasingly, we become so secure in our bubbles that we start accepting only information, whether it's true or not, that fits our opinions, instead of basing our opinions on the evidence that is out there. (Applause.)

And this trend represents a third threat to our democracy. But politics is a battle of ideas. That's how our democracy was designed. In the course of a healthy debate, we prioritize different goals, and the different means of reaching them. But without some common baseline of facts, without a willingness to admit new information, and concede that your opponent might be making a fair point, and that science and reason matter — (applause) — then we're going to keep talking past each other, and we'll make common ground and compromise impossible. (Applause.)

And isn't that part of what so often makes politics dispiriting? How can elected officials rage about deficits when we propose to spend money on preschool for kids, but not when we're cutting taxes for corporations? (Applause.) How do we excuse ethical lapses in our own party, but pounce when the other party does the same thing? It's not just dishonest, this selective sorting of the facts; it's self-defeating. Because, as my mother used to tell me, reality has a way of catching up with you. (Applause.)

Take the challenge of climate change. In just eight years, we've halved our dependence on foreign oil; we've doubled our renewable energy; we've led the world to an agreement that has the promise to save this planet. (Applause.) But without bolder action, our children won't have time to debate the existence of climate change. They'll be busy dealing with its effects: more environmental disasters, more economic disruptions, waves of climate refugees seeking sanctuary.

Now, we can and should argue about the best approach to solve the problem. But to simply deny the problem not only betrays future generations, it betrays the essential spirit of this country — the essential spirit of innovation and practical problem-solving that guided our Founders. (Applause.)

It is that spirit, born of the Enlightenment, that made us an economic powerhouse — the spirit that took flight at Kitty Hawk and Cape

Canaveral; the spirit that cures disease and put a computer in every pocket.

It's that spirit — a faith in reason, and enterprise, and the primacy of right over might — that allowed us to resist the lure of fascism and tyranny during the Great Depression; that allowed us to build a post-World War II order with other democracies, an order based not just on military power or national affiliations but built on principles — the rule of law, human rights, freedom of religion, and speech, and assembly, and an independent press. (Applause.)

That order is now being challenged — first by violent fanatics who claim to speak for Islam; more recently by autocrats in foreign capitals who see free markets and open democracies and civil society itself as a threat to their power. The peril each poses to our democracy is more far-reaching than a car bomb or a missile. It represents the fear of change; the fear of people who look or speak or pray differently; a contempt for the rule of law that holds leaders accountable; an intolerance of dissent and free thought; a belief that the sword or the gun or the bomb or the propaganda machine is the ultimate arbiter of what's true and what's right.

Because of the extraordinary courage of our men and women in uniform, because of our intelligence officers, and law enforcement, and diplomats who support our troops — (applause) — no foreign terrorist organization has successfully planned and executed an attack on our homeland these past eight years. (Applause.) And although Boston and Orlando and San Bernardino and Fort Hood remind us of how dangerous radicalization can be, our law enforcement agencies are more effective and vigilant than ever. We have taken out tens of thousands of terrorists — including bin Laden. (Applause.) The global coalition we're leading against ISIL has taken out their leaders, and taken away about half their territory. ISIL will be destroyed, and no one who threatens America will ever be safe. (Applause.)

And to all who serve or have served, it has been the honor of my lifetime to be your Commander-in-Chief. And we all owe you a deep debt of gratitude. (Applause.)

But protecting our way of life, that's not just the job of our military. Democracy can buckle when we give in to fear. So, just as we, as citizens, must remain vigilant against external aggression, we must guard against a weakening of the values that make us who we are. (Applause.)

And that's why, for the past eight years, I've worked to put the fight against terrorism on a firmer legal footing. That's why we've ended torture, worked to close Gitmo, reformed our laws governing surveillance to protect privacy and civil liberties. (Applause.) That's why I reject discrimination against Muslim Americans, who are just as patriotic as we are. (Applause.)

That's why we cannot withdraw from big global fights — to expand democracy, and human rights, and women's rights, and LGBT rights. No matter how imperfect our efforts, no matter how expedient ignoring such values may seem, that's part of defending America. For the fight against extremism and intolerance and sectarianism and chauvinism are of a piece with the fight against authoritarianism and nationalist aggression. If the scope of freedom and respect for the rule of law shrinks around the world, the likelihood of war within and between nations increases, and our own freedoms will eventually be threatened.

So let's be vigilant, but not afraid. (Applause.) ISIL will try to kill innocent people. But they cannot defeat America unless we betray our Constitution and our principles in the fight. (Applause.) Rivals like Russia or China cannot match our influence around the world — unless we give up what we stand for — (applause) — and turn ourselves into just another big country that bullies smaller neighbors.

Which brings me to my final point: Our democracy is threatened whenever we take it for granted. (Applause.) All of us, regardless of party, should be throwing ourselves into the task of rebuilding our democratic institutions. (Applause.) When voting rates in America are some of the lowest among advanced democracies, we should be making it easier, not harder, to vote. (Applause.) When trust in our institutions is low, we should reduce the corrosive influence of money in our politics, and insist on the principles of transparency and ethics in public service. (Applause.) When Congress is dysfunctional, we should draw our congressional districts to encourage politicians to cater to common sense and not rigid extremes. (Applause.)

But remember, none of this happens on its own. All of this depends on our participation; on each of us accepting the responsibility of citizenship, regardless of which way the pendulum of power happens to be swinging.

Our Constitution is a remarkable, beautiful gift. But it's really just a piece of parchment. It has no power on its own.

We, the people, give it power. (Applause.) We, the people, give it meaning. With our participation, and with the choices that we make, and the alliances that we forge. (Applause.) Whether or not we stand up for our freedoms. Whether or not we respect and enforce the rule of law. That's up to us. America is no fragile thing. But the gains of our long journey to freedom are not assured.

In his own farewell address, George Washington wrote that self-government is the underpinning of our safety, prosperity, and liberty, but "from different causes and from different quarters much pains will be taken...to weaken in your minds the conviction of this truth." And so we have to preserve this truth with "jealous anxiety;" that we should reject "the first dawning of every attempt to alienate any portion of our country from the rest or to enfeeble the sacred ties" that make us one. (Applause.)

America, we weaken those ties when we allow our political dialogue to become so corrosive that people of good character aren't even willing to enter into public service; so coarse with rancor that Americans with whom we disagree are seen not just as misguided but as malevolent. We weaken those ties when we define some of us as more American than others; when we write off the whole system as inevitably corrupt, and when we sit back and blame the leaders we elect without examining our own role in electing them. (Applause.)

It falls to each of us to be those anxious, jealous guardians of our democracy; to embrace the joyous task we've been given to continually try to improve this great nation of ours. Because for all our outward differences, we, in fact, all share the same proud title, the most important office in a democracy: Citizen. (Applause.) Citizen.

So, you see, that's what our democracy demands. It needs you. Not just when there's an election, not just when your own narrow interest is at stake, but over the full span of a lifetime. If you're tired of arguing with strangers on the Internet, try talking with one of them in real life. (Applause.) If something needs fixing, then lace up your shoes and do some organizing. (Applause.) If you're disappointed by your elected officials, grab a clipboard, get some signatures, and run for office yourself. (Applause.) Show up. Dive in. Stay at it.

Sometimes you'll win. Sometimes you'll lose. Presuming a reservoir of goodness in other people, that can be a risk, and there will be times when the process will disappoint you. But for those of us fortunate enough to have been a part of this work, and to see it up close, let me tell you, it can energize and inspire. And more often than not, your faith in America — and in Americans — will be confirmed. (Applause.)

Mine sure has been. Over the course of these eight years, I've seen the hopeful faces of young graduates and our newest military officers. I have mourned with grieving families searching for answers, and found grace in a Charleston church. I've seen our scientists help a paralyzed man regain his sense of touch. I've seen wounded warriors who at points were given up for dead walk again. I've seen our doctors and volunteers rebuild after earthquakes and stop pandemics in their tracks. I've seen the youngest of children remind us through their actions and through their generosity of our obligations to care for refugees, or work for peace, and, above all, to look out for each other. (Applause.)

So that faith that I placed all those years ago, not far from here, in the power of ordinary Americans to bring about change — that faith has been rewarded in ways I could not have possibly imagined. And I hope your faith has, too. Some of you here tonight or watching at home, you were there with us in 2004, in 2008, 2012 — (applause) — maybe you still can't believe we pulled this whole thing off. Let me tell you, you're not the only ones. (Laughter.)

Michelle — (applause) — Michelle LaVaughn Robinson, girl of the South Side — (applause) — for the past 25 years, you have not only been my wife and mother of my children, you have been my best friend. (Applause.) You took on a role you didn't ask for and you made it your own, with grace and with grit and with style and good humor. (Applause.) You made the White House a place that belongs to everybody.

(Applause.) And the new generation sets its sights higher because it has you as a role model. (Applause.) So you have made me proud. And you have made the country proud. (Applause.)

Malia and Sasha, under the strangest of circumstances, you have become two amazing young women. You are smart and you are beautiful, but more importantly, you are kind and you are thoughtful and you are full of passion. (Applause.) You wore the burden of years in the spotlight so easily. Of all that I've done in my life, I am most proud to be your dad. (Applause.)

To Joe Biden — (applause) — the scrappy kid from Scranton who became Delaware's favorite son — you were the first decision I made as a nominee, and it was the best. (Applause.) Not just because you have been a great Vice President, but because in the bargain, I gained a brother. And we love you and Jill like family, and your friendship has been one of the great joys of our lives. (Applause.)

To my remarkable staff: For eight years — and for some of you, a whole lot more — I have drawn from your energy, and every day I tried to reflect back what you displayed — heart, and character, and idealism. I've watched you grow up, get married, have kids, start incredible new journeys of your own. Even when times got tough and frustrating, you never let Washington get the better of you. You guarded against cynicism. And the only thing that makes me prouder than all the good that we've done is the thought of all the amazing things that you're going to achieve from here. (Applause.)

And to all of you out there — every organizer who moved to an unfamiliar town, every kind family who welcomed them in, every volunteer who knocked on doors, every young person who cast a ballot for the first time, every American who lived and breathed the hard work of change — you are the best supporters and organizers anybody could ever hope for, and I will be forever grateful. (Applause.) Because you did change the world. (Applause.) You did.

And that's why I leave this stage tonight even more optimistic about this country than when we started. Because I know our work has not only helped so many Americans, it has inspired so many Americans — especially so many young people out there — to believe that you can make a difference — (applause) — to hitch your wagon to something bigger than yourselves.

Let me tell you, this generation coming up — unselfish, altruistic, creative, patriotic — I've seen you in every corner of the country. You believe in a fair, and just, and inclusive America. (Applause.) You know that constant change has been America's hallmark; that it's not something to fear but something to embrace. You are willing to carry this hard work of democracy forward. You'll soon outnumber all of us, and I believe as a result the future is in good hands. (Applause.)

My fellow Americans, it has been the honor of my life to serve you. (Applause.) I won't stop. In fact, I will be right there with you, as a citizen, for all my remaining days. But for now, whether you are young or whether you're young at heart, I do have one final ask of you as your President — the same thing I asked when you took a chance on me eight years ago. I'm asking you to believe. Not in my ability to bring about change — but in yours.

I am asking you to hold fast to that faith written into our founding documents; that idea whispered by slaves and abolitionists; that spirit sung by immigrants and homesteaders and those who marched for justice; that creed reaffirmed by those who planted flags from foreign battlefields to the surface of the moon; a creed at the core of every American whose story is not yet written: Yes, we can. (Applause.)

Yes, we did. Yes, we can. (Applause.)

Thank you. God bless you. May God continue to bless the United States of America. (Applause.)

ACKNOWLEDGMENTS

THE PUBLISHERS WOULD like to thank the ownership and staff of the Harvard Book Store in Cambridge, Massachusetts, who first had the idea for this book, and whose manager, Carole Horne, gave us permission and encouragement to borrow that idea. The interior and cover design of this Melville House edition is based on the design of the bookstore's self-published local edition, which was by Mark Lamphier and Open All Night Books (cover) and Spencer Hawkes and Annie Bishai (interior).